Magnny Maisy de Barros Carvalho
Nara Caroline S. Rodrigues
Mariana Matos Arantes

# Ecological Soil-Cement Brick Incorporated with Babassu Coconut Fibre

Magnny Maisy de Barros Carvalho
Nara Caroline S. Rodrigues
Mariana Matos Arantes

# Ecological Soil-Cement Brick Incorporated with Babassu Coconut Fibre

ScienciaScripts

**Imprint**

Any brand names and product names mentioned in this book are subject to trademark, brand or patent protection and are trademarks or registered trademarks of their respective holders. The use of brand names, product names, common names, trade names, product descriptions etc. even without a particular marking in this work is in no way to be construed to mean that such names may be regarded as unrestricted in respect of trademark and brand protection legislation and could thus be used by anyone.

Cover image: www.ingimage.com

This book is a translation from the original published under ISBN 978-613-9-66757-4.

Publisher:
Sciencia Scripts
is a trademark of
Dodo Books Indian Ocean Ltd. and OmniScriptum S.R.L publishing group

120 High Road, East Finchley, London, N2 9ED, United Kingdom
Str. Armeneasca 28/1, office 1, Chisinau MD-2012, Republic of Moldova, Europe
Printed at: see last page
**ISBN: 978-620-8-06099-2**

I dedicate this work to my parents, Genivaldo and Magda, for all their efforts and encouragement to make this journey possible. Through them I was able to recognise that knowledge is an invaluable asset. To Maglly and my grandmother Veridiana for their support and companionship throughout these five years.

To my boyfriend André, for sharing every moment of this journey with me.

To these people I express all my affection and gratitude.

Magnny Maisy de Barros Carvalho


I dedicate this work to God, for having been my strength. To my dear parents, Zenaide and Lustosa, who never stopped trying to make my dreams come true. To my only and beloved sister, who always believed in my potential and encouraged me with irreplaceable advice. To my beloved grandmother, Rita, for all her affection and willingness to help me at any time.

You are everything to me.

Nara Caroline Silva Rodrigues

# ACKNOWLEDGEMENTS

We thank God for the gift of life and knowledge. For being our guide and our daily foundation.

To our family members, especially our parents, for their love, understanding and support, for pushing us to do our best throughout this academic journey.

To our friends and fellow students, for their support, companionship and for sharing unspeakable moments with us.

To the University Prayer Group, where we had a great time in communion with God and built beautiful friendships.

To our advisor, Mariana Matos Arantes, for her care, attention and for building this work with us; as well as being a great advisor, she has become a friend.

To the teachers who accompanied us, we thank you for all your teaching and guidance, through which we were able to grow professionally.

To the technicians, Irisvan and Marco Antônio, for all their support, patience and good times in the laboratory.

To CERMAR ceramics, on behalf of our classmate Renan Henrique, for giving us part of their raw material, the soil, which we used in this research.

To TOBASA, for providing us with the fibre from the epicarp of the babassu coconut used in this work.

To Mariane Femandes, my future work colleague, for her willingness and collaboration in making the bricks. And to all those who contributed in some way to making this work a reality.

"And they said to one another, 'Hey, let's make bricks and burn them well.

And their brick went for stone, and their bitumen for lime."

Genesis 11:3

# SUMMARY

In order to minimise environmental impact, the application of sustainable development has become substantial on the industrial scene. The combination of construction solutions, traditional materials and products sourced from the environment has been a solution to this reality. With this in mind, this work aims to incorporate natural fibres from the epicarp of the babassu coconut into ecological soil-cement bricks, which are already used in homes that aim to be economical and practical, using fibres from the epicarp of the babassu coconut. This material has been the livelihood of countless families in the North and Northeast and has been shown to have good mechanical and economic performance. The combination of these two materials has prompted several scientific research projects in the area of sustainable composites. In addition, the aim is to reduce waste disposal in the environment. Soil-cement bricks were made in ratios of 05:01 and 10:01 with the addition of fibre in masses of 0%, 1.5% and 3%. These bricks were evaluated at 28 days using water absorption and simple compression tests, as well as analysing the variation in specific mass. The results showed that the bricks with 3% fibre behaved better than the others, while the design that showed the best results was 05:01.

Keywords: Ecological brick. Cement soil. Babassu coconut fibre. Sustainable Development.

# SUMMARY

# CHAPTER 1

## INTRODUCTION

The system of chaos could lead to a consumerist society: Environmental degradation is creating a controversial scenario. Forests and rivers are replaced by desert environments where drought predominates. At other times, the natural imbalance causes floods and destruction due to the immense volume of water. Human beings are rampantly causing mass destruction and replacing the sustainability of trees with skyscrapers of buildings, which in the process of urban expansion inconsequentially degrade the soil and all of man's natural habitat.

As a consequence of this relentless pursuit of development, the climate is changing, which can be seen in the decline in air quality, the purity of river water and the amount of waste that is not easily degraded.

Introducing the concept of sustainability combined with development is no simple task, since the consumerist culture rooted in society operates in different social classes. However, there is a global discussion on the subject. The future is being shaped by the actions taken in the present, and man is already drawing up the answers.

Concerned about the future, governments and industries are gradually looking for ways to expand production, minimising environmental damage without compromising the quality of the products produced or services provided. The success of these sustainable development actions requires the engagement of various segments of society, with work that involves management that strives for quality, qualification and awareness of the employees involved in the production process, capable of discerning the importance of sustainable development from the most diverse sectors of the economy.

When the subject is brought up in the reality of civil engineering, construction is challenged by the quest to preserve nature. Quick results are demanded and economy is a priority factor, to the detriment of environmental defence. In this context, the construction processes used within the building industry are still considered backward when compared to the evolution of other branches, such as industry, because in addition to not being able to achieve continuously serialised production, there is a lack of technologies to automate the system and avoid the large disposal of materials.

It is in these failures that the construction industry constantly wastes time and money in its empirical way of carrying out services. Therefore, no matter how many attempts there are to minimise waste, the reduction in the cost of the work is negligible compared to the material that will be discarded.

With this in mind, the disposal and reuse of materials is no longer just a matter of economics, but an environmental necessity. It is with the ambition of making progress on inputs and tools that investment in research in this area has grown, since it is essential to find answers to sustainable development within the construction industry, whether it comes from new products or the adaptation and incorporation of existing materials.

In addition to research in the area, techniques that were considered outdated are being revived, as they are highly ecological and increasingly concerned about the inputs used. The use of ecological bricks is an old technique that has been reformulated and revamped and has been widely accepted by the public who want a clean building that is produced in a way that is friendly to the environment and has elements of sustainability rooted in its design.

As a result of this need, research is being carried out to explore composite materials in the hope of finding new performances and innovative properties. Along these lines, knowledge of the individual characteristics of each component of the composite is essential in order to achieve the desired performance of the final product.

In this context, in the search for natural reinforcements, the fibre extracted from the epicarp of the babassu coconut is seen as an innovative resource in the field of materials science. Its importance lies in the possession of certain characteristics such as: good tenacity, low density and dimensional solidity in its macromolecular structure.

Research has been carried out on the scientific scene with the intention of optimising the chances of achieving a composite with improved properties over those related to the conventional building material, the ecological soil-cement brick, and which meets the sustainable development plan that exists in the construction industry.

## 1.1    Objectives

### 1.1.1  General Objective

To verify the physical-mechanical influence of incorporating babassu coconut fibres into soil-cement bricks.

### 1.1.2  Specific objectives

- To analyse the mechanical behaviour of ecological bricks without babassu coconut fibre;
- Compare the different percentages of fibre incorporated into the brick so that the addition improves its deformation capacity;
- To compare the mechanical behaviour of the bricks with different proportions of cement added to the soil to make the brick;
- Evaluate the mechanical compressive strength of the composite developed;
- Measuring the specific weight of bricks with different fibre content variations;
- Identify the level of water absorption in traditional and fibre-embedded bricks.

## 1.2    Justification

Babassu is a type of palm found in abundance in certain Brazilian states, such as Tocantins, Maranhão and Piauí. Its main source of commercialisation is the oil from the kernels, which account for 7% of the fruit. The remaining 93 per cent of the fruit is the shell, made up of the epicarp, mesocarp and endocarp, which is usually discarded or sold at low prices due to the great difficulty in stocking it, as this material is light and takes up a large volume (EMBRAPA, 1984).

To prevent the bark from being discarded, there are a number of applications that can be made for this solid waste. One of these uses is as an energy source, in the form of firewood and charcoal.

Ecological soil-cement bricks are already being used in homes that are looking for savings, practicality and a contribution to the environment as they are not burnt like conventional bricks. It can also be left exposed, minimising the cost of the final finish.

The composite incorporated with the fibre from the babassu coconut has a reduced self-weight due to its low density. This reduction in the self-weight of the ecological brick will result in a significant reduction in the loads that the structure supports, helping to make overall savings in construction.

By adding fibre from babassu coconut shells to ecological bricks, the aim is to contribute to the development of research into sustainable materials, with a view to reducing waste disposed of in the environment, as well as making better use of it.

# CHAPTER 2

# HISTORICAL CONTEXT

The search for innovations that improve life has always been something that people have sought out. Since prehistoric times, there has been knowledge of actions and creations that bring more practicality and quality of life to everyday life.

Among the countless techniques that have been inherited by humankind, there is the application of ceramic material in masonry. This use arose from the need to build shelters made of durable and resistant materials that could withstand the elements.

Rocks, with all their grandeur and durability, served as a reference point, as they were noted for their support and safety. It was from these that men realised that by pressing raw earth they could obtain a similar, resistant material (LIMA *et al*, 2015).

In addition to the resistance observed, another valuable feature is the ease with which soil can be obtained in the wild. This means that it is not necessary to buy or transport it. On the other hand, this resource does not depend on industrial modification, which makes it economically viable and avoids environmental pollution (SILVA, 2000).

Bricks can be considered the oldest building material we have to date. Around 6000 years BC in Mesopotamia, bricks were made by hand. The clay was kneaded with straw, which gave the mixture solidity, then moulded with wood to obtain the desired shape and finally dried in the sun (KATO, 2002). The Mesopotamian peoples had easy access to clay, in a similar way to the Egyptians, extracting it from river valleys. However, approximately 3000 years BC, it was realised that by cooking in large ovens, the ability of the bricks to withstand the strain increased (REZENDE; GUILHERME; ALMEIDA, 2013).

In this vein, the brick became a substantial technological innovation in the field of construction and a major change in life, so that it influenced the transition from nomadic man living near caves to man needing fixed dwellings that provided comfort and security (PESTA et al, 2014). According to Pinto (2015), the Chinese Wall, as well as being a tourist attraction, is a clear example of the efficiency of this ancient construction technique. Its construction dates back 3000 years BC.

The use of bricks peaked at the beginning of the 17th century, when production became very popular. In 1985, the first machines for compressing earth are recorded, inspired by grape presses. The brick developed with its own production on site, as well as improvements in its shapes, dimensions and colours. They began to be used on a large scale for the infrastructure of factories, small bridges, warehouses, etc. (PESTA, 2014).

According to Pires (2004), with the advent of Portland cement in 1845, the method of construction using soil began to decline. Soil became a second-choice material and was preferentially used in rural areas. Studies show that one of the reasons was its reduced mechanical capacity and resistance to weathering, to the detriment of the new material on the market, Portland cement.

In Brazil, the use of earth to produce bricks is a legacy of the Portuguese colonisers (ARAÚJO, 2009). The Indians who lived in this hitherto unknown land already used clay to create ceramics and artefacts, but they didn't have as many techniques and tools to help them in their production processes.

The first technique used by the Portuguese to construct buildings in Brazil was rammed earth. As they needed their buildings to be durable, they opted for generous wall thicknesses. Based on these requirements, the process evolved to pau a pique (KATO, 2002).

The states that initially made the most use of technology in the colonial period were Minas Gerais, Mato Grosso, Goiás, Paraná and São Paulo, the latter being the one that most encouraged modernisation in this area (ARAÚJO, 2009). It was the change in São Paulo's building code in 1875 that restricted construction methods, banning ranches made of straw, grass or thatch. This encouraged the development of clay blocks (KATO, 2002).

According to Abiko (1983), the use of the soil cement construction method began in the United States in 1915, but research into the material was only carried out in 1935 by the American company Portland Cement Association. In Brazil, the material was first used in 1936. Inspired by the American experience, the Brazilian Portland Cement Association developed a dosage method for use in paving works (GRANDE, 2003). Since then, its use has been growing, as the material offers technical and economic advantages, especially in paving (LIMA, 2013).

# CHAPTER 3

# CONSTRUCTION MODELS IN CIVIL ENGINEERING

The definition of a construction model is basically linked to the construction stages in which they are formed. A variety of elements are linked together and complement the procedure as a whole. The goal is to achieve an acceptable end product (PASTRO, 2007).

## 3.1 Conventional Masonry

Conventional masonry is the traditional construction method most commonly used in Brazil. As shown in figure 1, the model consists of a "skeleton" made up of slabs, beams and walls that are filled with ceramic blocks, with no structural function, only sealing. There are no limits to subsequent remodelling, so it is known for developing creative freedom during execution.

Along these lines, masonry for sealing is used for both wall and wall construction. The advantage of its use lies in the flexibility to implement modifications and renovations. On the other hand, its use is disadvantageous due to the low productivity and abundant generation of solid waste during the electrical and plumbing installation stage after the walls have been built. It is therefore necessary to break the masonry to introduce the materials that make up the installations, which generates a lot of waste (FIGUEIREDO, 2015).

Figure 1. Conventional masonry construction Source: FIGUEIREDO, 2015.

## 3.2   Precast concrete system (reinforced concrete)

Concrete has been used since the time of the Greeks, back in the 5th century BC. During this period, mortar was produced from volcanic earth combined with lime, and was used in building infrastructure. However, the Greek civilisation had not yet discovered the structural function of concrete, so they were limited in overcoming large spans through structures (SOUZA, 2016).

Progress in the use of concrete in construction came about through the Roman civilisation, which developed advances through the knowledge acquired by the Greeks. Standards for managing the quality of materials were consolidated, which allowed for a chance to equalise all construction techniques. During this period, the Greeks used stone to fill the inside of walls, which is now known as cyclopean concrete. An example of Roman construction using concrete was the Pantheon in Rome in 126 AD (SOUZA, 2016).

In Brazil, the first signs of the use of reinforced concrete, once known as reinforced cement, were in buildings located in Copacabana under the supervision of engineer Carlos Poma. The company that carried out this construction was "Empresa de Construções Civis", which patented reinforced cement itself and later used it for other construction purposes. In terms of calculations for reinforced concrete structures, the first was for the bridge on the Maracanã River, in 1908 (PORTO; FERNANDES, 2015).

NBR 6118 (ABNT, 2014, p.21) defines reinforced concrete as: "Those whose structural behaviour depends on the adherence between concrete and reinforcement, and in which no initial elongation of the reinforcement is applied before the materialisation of this adherence."

It is also worth noting that concrete has a lower tensile strength, which is why it is supplemented with passive reinforcement to make up the reinforced concrete, as shown in figure 2. This construction method has many advantages, such as: high durability, little need for maintenance, ease of execution, abundant labour, resistance to bad weather and low cost during execution (PORTO; FERNANDES, 2015).

Figure 2 illustrates a house built using this construction technique.

Figure 2: Reinforced concrete walls Source: ABCP, 2010.

## 3.3 Structural masonry

This method is based on the use of blocks, which are inserted one on top of the other and bonded with mortar. The walls of this masonry act both as sealing elements and as structural elements that resist the building's loads. Masonry is not a new form of construction; it has been used in centuries past, albeit in a completely archaic way. In addition, bricks were commonly used for the pyramids of Egypt by the Assyrians and Persians in the construction of the Lighthouse of Alexandria and Notre Dame Cathedral (PASTRO, 2007).

Structural masonry, as shown in figure 3, is a modular production system that is entirely linked to current projects. Prior planning, for example, will allow installation mistakes or flow crossings to be identified. It is necessary to stick to the project in order to implement rationalisations that guarantee better use of materials, efficiency in the production stages and product quality. Compared to the conventional process, structural masonry has a more organised construction site due to the lower diversity of materials (KATO, 2002).

Figure 3. Structural masonry with concrete blocks
Source: ABCP, 2010.

# CHAPTER 4

# TYPES OF BRICK

## 4.1 Ceramic brick

The ceramics industry is known to be one of the oldest in the world. The first records in Brazil can be found at the end of the 19th century, when the Falchi pottery was set up.

Its popularisation intensified during the Industrial Revolution (BARBOSA, 2015), in which brick underwent various improvements and technological advances in its production. These include the improvement of rudimentary kilns to circular and tunnel-type kilns. And the pieces began to have standardised measurements and shapes (SANTOS, 2013).

In the ranking of the largest ceramics producers, Brazil stands out as one of the leading countries, after China, which produces around 3.5 billion $m^2$ . In the meantime, the ceramics industry has been considered a key component of the national economy.

Nowadays, demand in this industry is specifically for red ceramic products such as bricks, blocks, flooring and ceramic tiles, which are usually used in the finishing and cladding phases of construction work. It is therefore a potential generator of millions of jobs and income (VIEIRA, 2009).

The problem surrounding the use of the ceramics sector lies in the fact that manufacturing requires the firing of its products, which results in the release of gases that aggravate environmental degradation. The extraction of raw materials without proper restitution and the use of non-renewable sources are also factors that raise questions about the preference for this type of brick over other types.

The word Ceramic comes from the Greek *Keramike,* considered to be non-organic and non-metallic materials that are moulded in a plastic state between temperatures of 900 °C and 1,000 °C (VIEIRA, 2009). According to the Portuguese Ceramic Industry Association (APICER, 2000), ceramic bricks are classified according to their characteristics (solid, perforated and perforated) and application (face, filling and structural). In another classification, judged by INMETRO (2011), ceramic blocks can be arranged in two ways: they are called sealing when their function is only to make up the walls, as in figures 4 and 5:

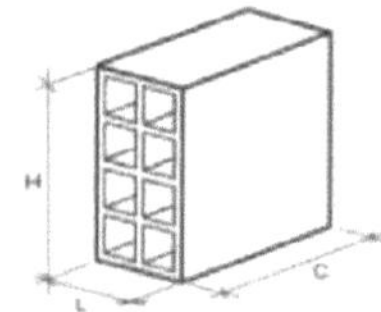 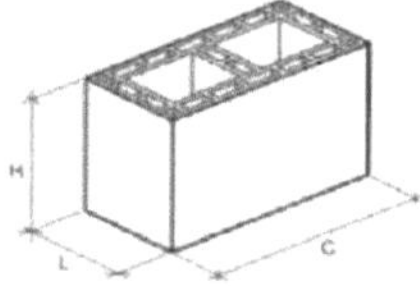

Figure 4 - Ceramic sealing block with horizontal holes, according to NBR 15270-1. Source: ABNT, 2005.
Figure 5 - Ceramic sealing block with vertical holes, according to NBR 15270-1. Source: ABNT, 2005.

Also according to INMETRO (2011), bricks that replace concrete pillars and beams are classified as structural, as in figures 6, 7 and 8.

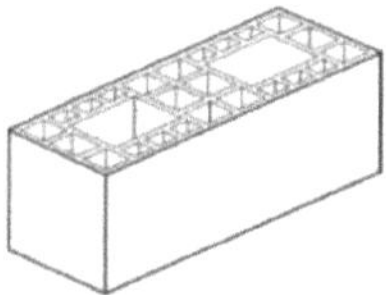 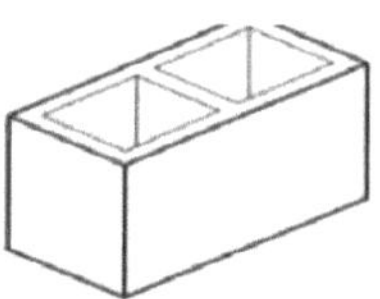

Figure 6 - Structural ceramic block for hollow walls, according to NBR 15270-2. Source: ABNT, 2005.
Figure 7. Structural block with solid walls (with solid internal walls), according to NBR 15270-2. Source: ABNT, 2005.

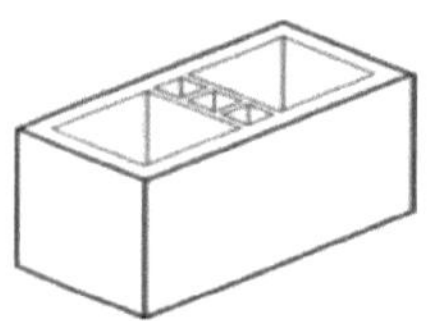

**Figura 8.   Structural ceramic block with solid walls (with hollow internal walls), according to NBR 15270 - 2.**

**Source: ABNT, 2005.**

**While the dimensions of both structural and sealing ceramic blocks are guided by NBR 15270 (ABNT, 2005) according to tables 1 and 2:**

Table 1. Manufacturing dimensions for structural ceramic blocks

| Dimensions (WxHxH) | | Manufacturing dimensions (cm) | | | | |
| Dimensional modulus M = 10 CM | Width (L) | Height (H) | Length C Main block | Length C 1/2 Block | Block L | Block T |
| --- | --- | --- | --- | --- | --- | --- |
| (5/4)M x (5/4)M x (5/2)M | | 11,5 | 24 | 11,5 | - | 36,5 |
| (5/4)M x (2)M x (5/2)M | 11,5 | 19 | 24 | 11,5 | - | 36,5 |
| (5/4)M x (2)M x (3)M | | | 29 | 14 | 26,5 | 41,5 |
| (5/4)M x (2)M x (4)M | | | 39 | 19 | 31,5 | 51,5 |
| (3/2)M x (2)M x (3)M | 14 | | 29 | 14 | - | 44 |
| (3/2)M x (2)M x (4)M | | 19 | 39 | 19 | 34 | 54 |
| (2)M x (2)M x (3)M | 19 | | 29 | 14 | 34 | 49 |
| (2)M x (2)M x (4)M | | | 39 | 19 | - | 59 |

L-block - block for tying into L-shaped walls
T-block - block for lashing to T-shaped walls
Source: AUTHOR (2017), adapted from NBR 1527-2 (ABNT, 2005)

Table 2. Manufacturing dimensions of ceramic sealing blocks

| Dimensions (WxHxH) | | Manufacturing dimensions (cm) | | |
| Dimensional modulus M = 10 CM | Width (L) | Height (H) | Length C Main block | Length C 1/2 Block |
| --- | --- | --- | --- | --- |
| (1)Mx(1)Mx(2)M | | 9 | 19 | 9 |
| (1)Mx(1)M x (5/2)M | | | 24 | 11,5 |
| (1)Mx(3/2)Mx(2)M | | 14 | 19 | 9 |
| (1)M x (3/2)M x (5/2)M | 9 | | 24 | 11,5 |
| (1)Mx(3/2)Mx(3)M | | | 29 | 14 |
| (1)Mx(2)Mx(2)M | | 19 | 19 | 9 |
| (1)M x (2)M x (5/2)M | | | 24 | 11,5 |
| (1)Mx(2)Mx(3)M | | | 29 | 14 |
| (1)Mx(2)Mx(4)M | | | 39 | 19 |
| (5/4)M x (5/4)M x (5/2)M | | 11,5 | 24 | 11,5 |
| (5/4)M x (3/2)M x (5/2)M | | 14 | 24 | 11,5 |
| (5/4)M x (2)M x (2)M | 11,5 | | 19 | 9 |
| (5/4)M x (2)M x (5/2)M | | 19 | 24 | 11,5 |
| (5/4)M x (2)M x (3)M | | | 29 | 14 |
| (5/4)M x (2)M x (4)M | | | 39 | 19 |

Source: AUTHOR (2017), adapted from NBR 15270 - 1 (ABNT, 2005)

With regard to the industrial process, Sousa (2003) states that production takes the form of preparing the raw materials, then forming them, drying them, baking them and then taking them out of the oven, where they go through a system that assesses their quality and, finally, the palletising process.

It is estimated that ceramic sealing blocks have good technical performance, including making it difficult for moisture to spread and providing thermal and acoustic insulation. Another feature to consider is their high resistance to wind pressure and rainwater infiltration. This material ensures that rooms can be adapted and divided, bringing safety and comfort to the occupants (BARBOSA, 2015).

On the other hand, there are difficulties in obtaining qualified labour and disposing of the various waste materials generated. It is necessary to ensure that the service is delivered in the shortest possible time, which leads to costly schedules or even poor execution of the work stages (BARBOSA, 2015).It is well known that the production process within the ceramics industry generates a lot of waste, such as defective and rejected parts that do not offer the quality and performance desired by the producer. Along these lines, this problem is part of an environmental approach that aims to prevent these materials from being transported to landfill sites (VIEIRA, 2009).

## 4.2  Concrete block

Known as concrete block, concrete brick or even structural block, the variation in names defines one of its main characteristics: being used in structural masonry.Concrete block is defined by Medeiros (1993) as "the masonry unit made up of a homogeneous mixture, suitably proportioned, of Portland cement, fine and coarse aggregate, formed through vibration and pressing, with dimensions and heights greater than 250x120x55 mm". NBR 6136 (2016) standardises these dimensions, as shown in table 3:

Table 3. Standardised concrete block dimensions

| Nominal dimensions (cm) | Name | Standardised dimensions (mm) | | |
|---|---|---|---|---|
| | | Width | Height | Length |
| 20 x 20 x 40 | M-20 | 190 | 190 | 390 |
| 20 x 20 x 20 | | 190 | 190 | 190 |
| 15x20x40 | M-15 | 140 | 190 | 390 |
| 15X20X20 | | 140 | 190 | 190 |

Source: AUTHOR (2017), adapted from NBR 6136 (ABNT, 2016)

Incidentally, the use of concrete blocks in masonry began shortly after the advent of Portland cement, with the production of large, solid pieces of concrete. From then on, there was a need to modernise the production process, but today the materials and dosing procedures remain basically the same (FILHO, 2007).

As it is used in structural masonry, it is also a determinant of the main characteristics of this system in terms of the resistance of the structural elements. It can be found in various sizes, shapes, patterns, textures and colours (IZQUIERDO, 2011). Figure 9 shows some of the main concrete block profiles used and their dimensions:

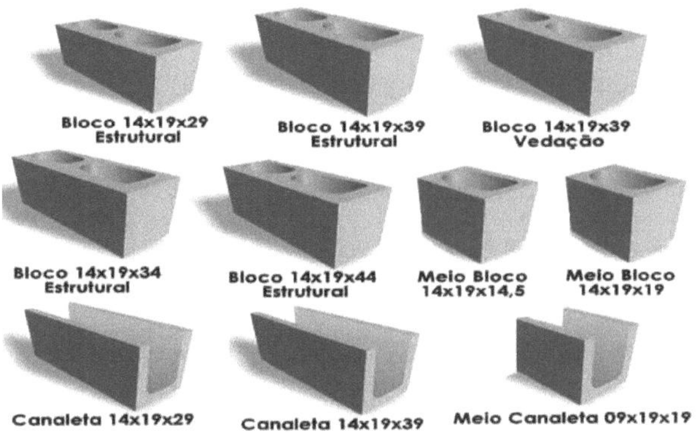


Figure 9. Precast concrete
Source: GOLDEN BLOCOS DE CONCRETO, 2017.

In terms of physical and mechanical characteristics, the main masonry design parameter is the block's compressive strength. The adherence of the block to the mortar is directly defined by the initial water absorption capacity. Total absorption indicates the amount of voids and allows the density of the block to be determined (MEDEIROS, 1993).

It should also be added that only by using blocks with little variation in their dimensions is it possible to obtain walls with a defined geometry. This characteristic makes it easier for the bricklayer to lay the blocks and influences the masonry's good structural performance (MEDEIROS, 1993).

## 4.3    Ecological brick

Ecological bricks are bricks that are not burnt during manufacture and therefore reduce pollution of the ecosystem.

It's worth noting that sustainable bricks, when compared to conventional bricks, have two holes to help with electrical and plumbing installations. It also has the advantage of providing both thermal and acoustic comfort (MOTTA *et al*, 2014). To better understand the functionality and history of its use, other types of bricks that also use raw earth will be analysed.

### 4.3.1  Taipa de Pilão

Taipa de pilão (rammed earth) is an ancient construction model that was widely used to build churches, especially during the colonisation of Brazil between the 16th and 14th centuries. Today, there are records of rammed earth houses that are 300 years old (SILVA, 2000). It was once considered one of the most solid systems within the raw earth construction sector, given that the walls were built monolithically and solidified progressively over the years. In addition, taipa de pilão helps to regulate the temperature and humidity inside the building (SILVA, 2000). Figure 10 illustrates a house of this type of construction technique, located in Butantã, São Paulo.

Figure 10. House with rammed earth walls Source: PROMPT, 2008.

The construction process is based on the horizontal compression of the clay into 15 cm high layers that are compacted in wooden moulds, the so-called "taipas". This process guarantees uniform compaction.

This technique is carried out with the aid of a pestle (see figure 11), where the soil is initially slightly moistened. The tools used to carry it out are usually wood, nails and soil (NASCIMENTO, 2013).

As the building is made of raw earth, it is always advisable to use plastic sheeting during the production stages. This is related to exposure to rain, should it occur. Avoiding possible damage to the work.

Figure 11. Tilling process. Source: MINKE, 2005.

The tilling technique is clearly explained by Zorowich (2014):

> The layers of clay are reduced to half their height by the process of piling. When the piled earth reaches about 2/3 of the height of the taipal, it is transversely covered with small round sticks wrapped in leaves, usually from banana trees, producing cylindrical holes called "cabodás" that allow the taipal to be anchored in a new position (ZOROWICH, 2014).

The quality of this construction process depends primarily on the choice of soil, as well as its dosage and compaction. It is therefore necessary to select soil samples in the area you want to build and, in this process, discard the first 30 centimetres of soil, as it is fertile and may contain organic material, which could compromise the material's resistance (MONTORO, 1994).

4.3.2  Adobe

The term "Adobe" has Arabic origins and means raw brick. This construction technique, which is still used in Brazil in various regions, is known for its simplicity and affordability. It consists of mixing earth and water in such a way as to achieve a flexible consistency. There is no use of cement in the mixture, much less the firing or baking of the block. So, as soon as it is moulded, it dries naturally. The moulds come in a variety of sizes in Brazil, where the material is available in 8x12x25 cm and 7x15x31 cm (ALVARENGA, 1990).

To make the bricks, see figure 12. The mortar is made from clay and sand, and is mixed with vegetable fibres such as straw where possible.

Knead the clay with your feet and once it has reached the ideal consistency, leave it to rest for 2 days, protected so that it doesn't get wet from rainwater. After 2 days, the clay should be beaten again and placed in the wooden mould, without first wetting it so that the wood doesn't suck the water out of the dough. Level the mixture with a ruler, then carefully remove the brick from the mould so as not to break it. If the brick deforms, then the mixture is too wet, add clay, if the brick cracks, then the mixture is too dry, add water. Sand and straw should also be added to the mix in both cases (SILVA, 2000).

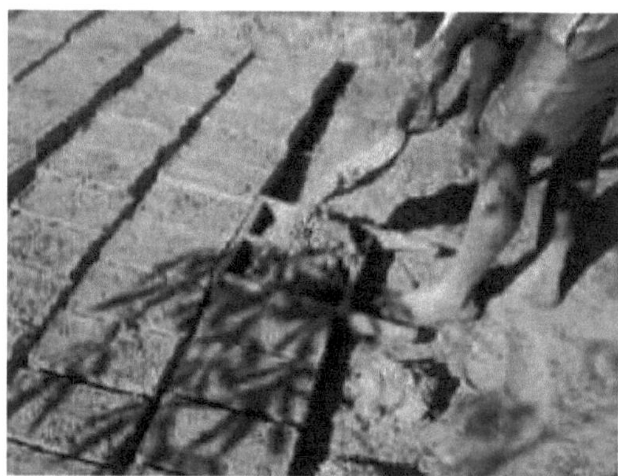

Figure 12. Manufacturing process for adobe bricks.
Source: ALEXANDRIA; LOPES, 2006.

The laying process is carried out with the clay itself still damp and does not require the use of mortar. When properly constructed, adobe buildings can have a long lifespan. Despite the widespread prejudice against its use, this system has often been revived and preserved. The problem with this construction method is the appearance of cracks and fissures after shrinkage, usually in cases of high sun exposure. On the other hand, this complication can be minimised by adding branches and twigs between adobe connections (SANTOS, 2004).

4.3.3Soil    cement

Ecological bricks are gaining market share because they are a construction technique that generates little environmental impact compared to other more modern techniques.

According to Motta et *al* (2014), the term "ecological brick" is linked to the fact that it is manufactured without burning wood and fuel, thus avoiding atmospheric pollution and deforestation. It is a compound formed from mixing soil with cement, which is then pressed together. Its shape may or may not contain holes that allow for electrical and plumbing installations, eliminating the need to break down walls. Instead of conventional adhesive mortar, a special glue is used to lay it. It is known for being a modular system that develops a uniform masonry, which reduces plastering losses. If it fulfils the established strengths within the project criteria, it can be used for sealing or structural masonry (SALA, 2006).

Figure 13. Soil cement brick masonry Source: FURTADO, 2007.

The brick is usually manufactured on site and takes longer than conventional bricks (MORAIS, 2014). As for the criteria for choosing the soil, it must fulfil certain predefined characteristics. The dimensions of the ecological brick are shown in tables 4 and 5.

Table 4. Soil selection criteria

| Features | Requirements (%) |
|---|---|
| % of soil passing the 4.8 mm sieve (No. 4) | 100 |
| % of soil passing the 0.075 mm sieve (no. 200) | 10-50 |
| Liquidity limit | ≤45 |
| Plasticity limit | ≤ 18 |

Source: ABCP, 1985.

Table 5: Types and dimensions (in mm) of soil-cement bricks, according to NBR 8492.

| Types | Length | Width | Height |
|---|---|---|---|
| Massive | 200 | 100 | 50 |
| With holes | 240 | 120 | 70 |

Source: ABNT, 2012.

Once they have been pressed, they have to go through a curing process, moistened, in order to acquire resistance. As it is a simple process, it does not require skilled labour. Even if manufacturing is not based on any specific technical standard, its procedures must comply with the standards of the Brazilian Association of Technical Standards (ABNT) (MOTTA *et al*, 2014).

The revival of this modular system brings several benefits, in addition to environmental ones. These include cost savings and reduced energy and water consumption. According to Motta et *al* (2014), ecological brickwork achieved a cost reduction of around 21% compared to conventional brickwork, with a comparative cost of R$ 21.047/m$^2$ against R$ 26.59/m$^2$ . This makes it an excellent option for building low-income housing.

Another point to add is the thermal and acoustic comfort that buildings made of this type of brick have, compared to those made of ceramic bricks, for example. This is because raw earth is a poor conductor of heat. It also provides good impermeability and resistance, which ensures greater durability and reduced maintenance (OLIVEIRA, 2011).

Soil-cement brick buildings do not require plastering, rendering or rendering, as they have a smooth finish due to the pressing of their faces. Therefore, only painting would be necessary for a good finish that provides hygiene, comfort and permeability to the walls (OLIVEIRA, 2011).

It's worth noting that according to the Micro and Small Business Support Service (SEBRAE) (2017), 1/3 of the materials used in conventional construction are discarded, while losses are much lower in building systems using soil cement bricks. Other advantages of using soil cement bricks over conventional systems include a 30 per cent reduction in construction time, as well as savings of 50 per cent on iron and 70 per cent on mortar and concrete.

According to Oliveira (2011), the main disadvantage of establishing this material is the constant need to characterise the soil at the extraction site, since it is so varied in nature. The tests for its evaluation are: simple compression, compaction and granulometry. Dosages, when not carried out correctly, can lead to erosion and pathologies.
Likewise, the unbridled use of this product can intensify erosive actions in the ecosystem (MOTTA *et al,* 2014).

Society still has a prejudiced attitude when faced with this technique, since the current trend is to move away from ancestral knowledge and follow industrialisation concepts and technological innovations. In other words, ecological brick construction faces prejudice and rejection due to ignorance of the benefits it provides. The use of ecological bricks provides healthy and bioclimatic conditions as long as they are properly treated and cared for in every building (SILVA, 2000).

# CHAPTER 5

**PLANT FIBRES**

Faced with a scenario that values sustainability, the incentive to use natural fibres as an addition to composites is intensifying and the number of research projects in this area is growing considerably. Thus, the idea of adding plant waste to fragile matrices is an alternative aimed at reducing plant waste. Most of this waste is characterised as rejects due to its uselessness and inadequate destination, and could be put to better use in the creation of new, less polluting materials (SILVA, 2005).

Although this reality has been expanding in recent decades, this technique is not unknown. Plant fibres have been used for millennia to reinforce materials. Among the most commonly used natural fibres are coconut, mallow, bamboo, sisal, jute, piassava, flax, sugar cane and cellulose (SPECHT, 2000).

According to Silva (2005), plant fibres are bundles made up of individual cells composed of microfibrils, which are rich in cellulose. The numerous cells that make up the fibre are grouped together by the intercellular lamella made up of amorphous substances such as hemicellulose, pectin and lignin. Fibres are characterised by their physical and mechanical properties, such as void volume and water absorption. The volume of voids influences the high absorption in the first moments of immersion, which is a negative point in the water/binder ratio of the matrix, the swelling of the material and subsequent shrinkage. However, the high number of voids helps to reduce the specific mass and thermal conductivity, as well as increasing acoustic absorption.

Santos (2004) states that the highest quality of fibre-reinforced materials is found in the state after cracking. It is at this stage that the added fibres contribute by increasing the material's energy absorption capacity, resulting in greater resistance to traction, impact and fatigue.

It is important to note that the improvements resulting from the addition of a fibre to the soil vary according to the characteristics of the fibre, the characteristics of the soil, the confinement tension and the form of loading (SILVA, 2005). Likewise, another factor that must be estimated is knowledge of the relationship between the matrix and the composite, as well as the contribution of each to the final result. In short, it is with these analyses that you get a material with the characteristics you expect (SANTOS, 2004).

## 5.1 Babassu Coconut Vegetable Fibre

Also known as coco-macaco, baguaçu, pindoba, palha branca or various other names, the babaçu, shown in Figure 14, is a palm of the Arecaceae family that can be found in several South American countries (CARRAZA; SILVA; ÁVILA, 2012). In Brazil, according to the Brazilian Agricultural Research Corporation (EMBRAPA) (1984), babaçuais are found in the Northeast, North and Centre West regions, as well as in the state of Minas Gerais. In addition, the Northeast region has the highest production of almonds and the largest area occupied by coconut plantations.

Figure 14. Babassu bunch in front and palm trees of the species behind.
Source: CERRATINGA, 2017.

In addition, this palm has a simple trunk, but it is large and imposing and can reach a height of up to 20 metres. The leaves are compound, green, arched and measure around 8 metres in length. This plant produces brown fruit with edible oil seeds. There are four bunches per palm, with an average of 150 to 250 coconuts per bunch (FRANCO, 2010).

In its vast dissemination throughout Brazil, babassu is found in different types of soil and alternates with primitive or derived vegetation, such as forest, savannah, pasture and plough. Due to the diversity of climate, vegetation and soil to which it is exposed and considering its invasive power in plant succession, it is recommended to study and research the various behaviours of this palm in different environments in order to understand its variations (EMBRAPA, 1984).

The coconut produced by babassu falls spontaneously after it reaches maturity. The fruit is broken, mostly by women, using only an axe and a piece of wood. The productivity of this rudimentary system is low and requires care.
to avoid accidents. The coconut can be broken in the babassu grove itself, but when the forest is closed it is broken in the houses of the breakers or in centralised areas where it is transported. There are now mechanical breaking machines, which are gradually replacing the manual process.

The babassu fruit, illustrated in figure 15, is made up of four usable parts. These are: the epicarp (11%), the mesocarp (23%), the endocarp (59%) and the kernels (7%). Even though it is entirely usable, the coconut shell (a combination of the epicarp, mesocarp and endocarp, which makes up 93% of the fruit) is discarded by the breakers (EMBRAPA, 1984).

Figure 15. Transverse and longitudinal section of babassu coconut.
Source: Author, 2017.

Most babassu groves are used to produce almonds. Almonds are good quality oilseeds and have similar characteristics to coconut oil. An average of 3 to 4 kernels can be found per coconut. They are used in the food, cosmetics, veterinary, pharmaceutical and chemical industries (FRANCO, 2010).

The endocarp is the most resistant layer of the babassu coconut and represents more than half of the fruit. It is used as charcoal because it is of excellent quality, superior to wood charcoal, and can also be used to make various handicrafts.

The mesocarp is the layer that lies between the epicarp and the endocarp. As it is rich in starch, it is used to produce "in natura" starch flour, starch, pregelatinised starch, glucose and ethyl alcohol (EMBRAPA, 1984).

The epicarp, the part of the coconut used in this study, is the outer layer and is rigid and fibrous. It can be used as tree fern, to produce plant pots, upholstery, packaging to replace Styrofoam, organic fertiliser and charcoal, and its burning characteristics are similar to the endocarp (CARRAZA; SILVA; ÁVILA, 2012).

Even though babassu coconut shells have a variety of applications, they are still discarded in large quantities. Due to factors such as its large volume, which makes it difficult to store, coconut breakers choose to discard this material instead of looking for companies that can use it.

It is important to be aware of the difficulties that companies have in utilising all parts of the babassu tree, as well as the performance of industries and the potential market in the region, in order to define the best areas for babassu application (EMBRAPA, 1984).

# CHAPTER 6

## COMPOSITE MATERIALS

Composite materials are defined as combining at least two materials, which even after mixing can be identified, and are therefore heterogeneous. They usually have two phases: a matrix phase and a reinforcement phase.

The matrix phase is responsible for transferring and homogenising the stresses that the fibre phase supports. The reinforcement phase is responsible for being the element that reinforces and acts as a bridge for the transfer of efforts (IZQUIERDO, 2011).

According to Loos (2014), the matrix is a continuous phase that shapes the structure. As the continuous phase, it surrounds and covers the reinforcement, as well as being the phase that is directly exposed to the environment.

It is the matrix that first comes into contact with any force exerted on the composite, but it is not the strongest part of the composite, since it is only expected to transfer all these applied loads to the reinforcing phase. As for the reinforcing phase, these are fibres that can be continuously arranged, long or short. They can also be orientated or not.

The size of the fibres influences the way the composite breaks, as shown in figure 16. While the material without fibres would break entirely in the direction of the failure, the material with short fibres would form small cracks on the surface. The composite with long fibres would prevent the failure from propagating throughout the material.

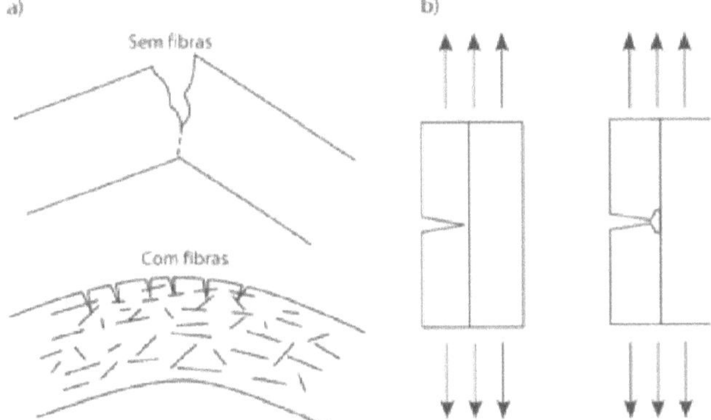


Figure 16. Representation of fault propagation being interrupted: (a) Composite with short fibres and (b) Composite with long fibres.

Source: LOOS, 2014.

Composites can be classified into three distinct classes: particle-loaded composites, discontinuous fibre composites and continuous fibre composites (PANZERA, 2007), as illustrated in figure 17:

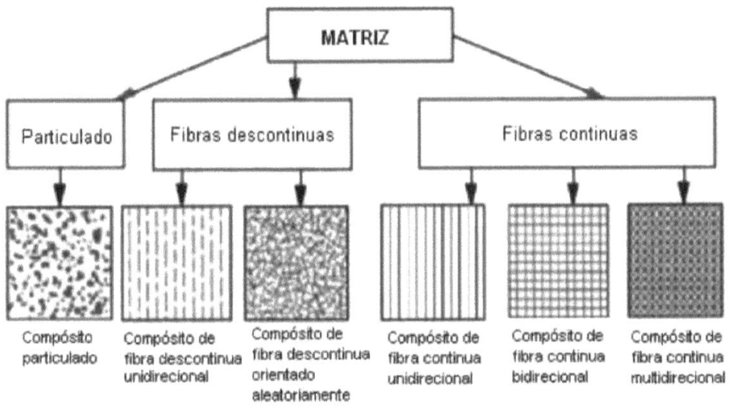


Figura 17. Main types of composites.

Source: PANZERA, 2007.

The greater the amount of fibres, the better the strength and toughness of the material, as the number of micro-cracks the fibres will act on will be greater, absorbing part of the internal tensions. However, an increase in the number of fibres also reduces the workability of the mixture, which can hinder the homogeneity procedure (IZQUIERDO, 2011).

## 17.1  Ceramic matrix composite with natural fibres

Ceramic materials are resistant to oxidation and deterioration even at high temperatures, but because of their brittle fracture, they are not viable for some applications. However, this fracture toughness has been considerably improved by research into new composites whose matrix is a ceramic material. These composites tend to resist fracture through reinforcement, which can be in particulate or fibre form (PANZERA, 2007).

Ceramic materials are made up of metallic and non-metallic elements formed through the action of heat and subsequent cooling. In this case, traditional ceramic materials will be considered, those formed from clay raw materials, above all ceramic material formed from soil and cement.

The technique of building with soil or soil-cement, although old, is still lacking in studies. Due to the great durability shown in old buildings that use these materials, there is a need to improve and invest in the sector (SILVA, 2005).

When studying composites, the use of soils, cements and fibres in their production is noteworthy. The use of recyclable materials or materials of natural origin is growing considerably because they are materials that do not destroy the environment, combining the concept of development with the concept of sustainability (SILVA, 2005).

Another point to note is that natural fibres are an excellent application in fragile matrices due to their abundance, low cost and low energy consumption to produce them.

Increasing the fibre content also increases the compressive strength value. Oliveira (2011) explains that this is due to various factors such as: the transfer of loads from the fibre to the soil, redistributing tensions and improving the interactions of the matrix-fibre interfaces. It also favours hydraulic conductivity, as there is an increase in permeability, thus making cement hydration reactions more effective. Another factor observed was that the composite did not break abruptly, continuing with the two crack faces joined after the matrix broke in the compression test.

The use of composites with ceramic material and vegetable fibre additions, such as soil cement bricks with added babassu coconut fibre, is viable due to the ease with which materials can be acquired, especially in the north and northeast of Brazil. It also has the socio-environmental function of an ecological brick and fulfils the requirements of sustainable development.

# CHAPTER 7

## MATERIAL AND METHODS

The test specimens were prepared using the following materials: soil, cement, fibre and water. The soil had proportions of coarse sand from a riverbed and clay donated by a ceramics company in the city of Araguaína-TO. The cement used to make the bricks was Nassau brand CPIV-32 (Pozzolanic Portland Cement). The babassu epicarp fibres came from a company that works with the total use of babassu coconut in the city of Tocantinópolis-TO. The water was supplied by the city of Araguaína-TO's water supply system, as it is characterised as potable and therefore suitable for such use.

### 17.2  Soil characterisation

In order to comply with the requirements set out in NBR 8492 (ABNT, 2012), which establishes the tests for dimensional analysis, determination of compressive strength and water absorption, the soil used was made up of two different soils: coarse sand and clay. For this composition, 70% sandy soil and 30% clay soil were mixed. The soil samples were subjected to preliminary tests for better classification and to determine the mix to be used to make the soil cement brick.

Three (3) tests were carried out: Particle size analysis, consistency indices and specific mass of the grains.

7.1.1 Particle size analysis

The particle size distribution test consists of determining the size of the various particles in the soil and the relative proportions in which they are found.

It was carried out in accordance with NBR 7181 (ABNT, 2016). According to the standard, this test is divided into three stages: coarse sieving, sedimentation and fine sieving.

The material was prepared as specified in NBR 6457 (ABNT, 2016), after which it was passed through a sieve with an opening of 2mm, mesh number #10, so that, in accordance with the standard, the coarse sieving of the retained material could be carried out. As it is not desirable for the soil sample from the brick to exceed a particle size of 4.8mm, only the material retained in the 2mm mesh is weighed. With the material passing through the sieve, the sedimentation test and fine sieving are carried out.

The sedimentation technique helps to determine the fine material in the soil sample. The soil used is characterised as sand, so to carry out the test, 120g of the material passing through the 2mm sieve is weighed, placed in a beaker and added to a solution made up of distilled water and a dispersing agent, in this case sodium hexametaphosphate, as shown in figure 18. The mixture was left to stand for 12 hours.

Figure 18. Soil sample with distilled water and sodium hexametaphosphate.
Source: AUTHOR, 2017.

After this period, the material was transferred to the dispersion beaker with distilled water, where it was stirred for 15 minutes. The material was then transferred to a 1000ml beaker and filled with distilled water up to the reference mark. The beaker was sealed with PVC film to prevent the sample from leaking during stirring.

This was done with the help of one hand. The process consisted of covering the mouth of the cylinder and shaking the contents for around 1 minute, using energetic rotational movements. Once this stage was completed, the cylinder was placed on a bench, where the density and then the temperature of the soil were measured, as shown in figure 19.

Figure 19. Test tube with the soil for the sedimentation test.

Source: AUTHOR, 2017.

The thermometer and densimeter readings were taken at 30s, 1min and 2 min, 4min, 8 min, 15 min, 30 min, 1 h, 2h, 4h, 8h and 24h, measured according to the start of the measurements.

The following formula was used to calculate the percentages corresponding to each densimeter reading:

$$Qs = N \ x \ \frac{\rho_s}{\rho_s - \rho_w} \ x \ \frac{V \ x \ \rho_w \ x \ (L - L_d)}{\frac{M_h}{100+w} \ x \ 100}$$

(Eq. 1)

Where:

Qs= percentage of soil in suspension at the moment the densimeter is read;

N = percentage of material passing the 2.0 mm sieve, calculated as indicated above;

v = volume of the suspension, in cm;

L = Densimeter reading in the suspension;

Ld= reading of the densimeter in the dispersing medium, at the same temperature as the suspension;

Mh= mass of wet material subjected to sedimentation, in g;

w = hygroscopic moisture of the material passed through the 2.0 mm sieve.

To calculate the maximum diameter of the suspension particle, the formula is:

$$d = \sqrt{\frac{1800 \; x \; \eta}{\rho_s - \rho_w}} \; x \; \frac{a}{t}$$

(Eq. 2)

Where:

d = maximum particle diameter in mm;

n = coefficient of viscosity of the dispersing medium at the test temperature, in g x s/cm2;

a = height at which the particles fall.

After carrying out the sedimentation test, the soil in the beaker was poured through a 0.075mm sieve and washed with drinking water at low pressure. The material retained on the sieve after this process was taken to the oven and used in the fine sieving test process.

To carry out the fine sieving, the sample removed from the oven is placed on a series of sieves with apertures of 1.2, 0.6, 0.42, 0.25, 0.15 and 0.075mm, corresponding to mesh numbers 16, 30, 40, 60, 100 and 200 (figure 20), with the aid of a sieve shaker.

Figure 20. Set of sieves for fine screening.
Source: AUTHOR, 2017.

The material retained on each sieve is weighed and recorded, and from these values the particle size curve is formed.

The following equation was used to calculate the percentage of material passing through the sieves in fine screening:

$$Qf = \frac{M_h \times 100 \quad M_i \times (100+w)}{M_h \times 100} \; x \; N$$

(Eq. 3)

Where:

Qf= percentage of material passed through each sieve;

Mh= mass of wet material subjected to sedimentation;

Mi= mass of retained material accumulated in each sieve;

N = percentage of material that passes the 2.0 mm sieve.

In order to obtain precision in the particle size curve drawn using this test, it is necessary to know the specific mass of the soil.

7.1.2 Specific mass of the grains

Equivalent to the average value of the specific mass of each soil particle contained in the sample. The test was carried out using a pycnometer for both samples, based on NBR 6459 (ABNT, 1984).

It was necessary to weigh out 10g of dry soil and submerge it in distilled water. For 24 hours. The empty pycnometer was then weighed and the soil sample and distilled water added. This container was heated on a metal plate for 15 minutes to expel the air present between the grains of soil. It was then allowed to cool until it reached room temperature. It was then filled with distilled water up to the meniscus, where the temperature of the mixture was measured. This measurement was noted down.

The pycnometer was then wiped dry and weighed.

The same process was repeated with another soil sample.

Thus, the real density of the soil is calculated on the basis of the pycnometer weighing values and their variations in content, Equation 4.

$$\delta_s = \frac{\dfrac{100 \times M_1}{100 + w}}{\dfrac{100 \times M_1}{100 + w} + M_3 - M_2} \delta_t$$

(Eq. 4)

Where:

$\delta_s$ = Specific weight of soil grains;

$\delta_t$ = Density of water;

Mi= Weight of empty pycnometer;

M2= Weight of pycnometer plus sample;

Ma= Weight of pycnometer plus sample plus water;

w= soil moisture.

### 7.1.3  consistency indices

As the name suggests, the consistency indices or Atterberg limits (liquidity limit and plasticity limit of soils) are values that express the soil's ability to be moulded, under a certain humidity condition and without volume variation. It is determined using two tests: Liquidity Limit, governed by NBR 6459 (ABNT, 2016), and Plasticity Limit, governed by NBR 7180 (ABNT, 2016).

The formula for determining the consistency index is:

$$IC = \frac{LL-h}{LL-L}$$

(Eq. 5)

Where:

CI= consistency index

LL= Liquidity Limit

h= field moisture content

LP= Plasticity Limit

To determine the liquidity limit, the part of the sample that passed through the 40 mesh sieve, i.e. 0.42mm, was separated and weighed. This material was transferred to a glass dish, where water was gradually added until the mass was homogenised.

After mixing the material, it was transferred to the shell of the Casagrande apparatus. This apparatus consists of a semi-spherical shell that moves towards a rubber base by means of a

gear and crank mechanism. A cutter or chisel is used next to the device to divide the analysed soil into two portions.

With the soil laid out on the shell, a centre line was made (figure 21) so that the opening was 1cm thick. A groove was then made in the middle of the mass, using a chisel, in the direction of the length of the device. The crank was turned at a rate of two revolutions per second, counting the number of strokes needed for the lower edges of the grooves to join along the entire length of the device.

Figure 21. Homogenisation of the soil for the liquidity limit test.
Source: AUTHOR, 2017.

A small amount of the material was removed where the edges of the groove met and placed in the aluminium capsule to determine the moisture content. The capsule and the soil were weighed. The whole set was then taken to the oven and weighed again after 24 hours.

All the remaining material from the test was transferred to the glass plate, where more water was added. The procedure was repeated four more times to obtain the shank bond at 10, 14, 20, 33 and 45 blows.

The liquidity limit was determined by the moisture content corresponding to the intersection of 25 blows. To find this value, a scatter plot is drawn, obtained by dividing the number of blows found in the test by the moisture content of the capsules for each blow. The moisture content is calculated using Equation 6.

$$Hll = \frac{Ph - Ps}{Ph} * 100$$

(Eq. 6)

Where:

Hll = Soil moisture content in %;

Ph = Gross wet weight, in g;

$P_s$ = Dry weight, in g.

Figure 22. Liquidity limit test.
Source: AUTHOR, 2017.

To determine the plasticity limit test, the soil was formed on a glass plate, where distilled water was added in small portions. The material was kneaded with a spatula until a homogeneous paste with a plastic consistency was obtained. 10g of the sample was used to form a small ball, which was moulded against the glass plate with pressure to form a cylinder, figure 23. This cylinder should have a diameter of 3 mm, the size of an auxiliary template, before fragmenting.

Figure 23. Plasticity test. Source: AUTHOR, 2017.

Once fragmented, the fragments were collected to determine their moisture content. The fragments were taken to the oven for 12 hours to obtain their dry weight. With the average of the moisture values of the samples, it is finally possible to obtain the moisture values of the sample, using Equation 7.

$$Hlp = \left(\frac{(Ph-Ps)}{Ph}\right) * 100$$

(Eq. 7)

Where:

Hlp = Soil moisture content in %;

Ph= Gross wet weight, in g;

$P_s$ = Dry weight, in g.

## 7.2 Characterisation of babassu coconut fibre

Because there were impurities such as earth, ash and other parts of the babassu coconut next to the fibres, it was necessary to wash all the material. After washing and drying in an oven for 24 hours at 60°, as shown in figure 24, some tests were carried out to better identify the characteristics of this fibre: Specific Mass, Moisture Content and Water Absorption.

Figure 24. Babassu coconut epicarp fibre after washing.
Source: AUTHOR, 2017.

### 7.2.1 Specific mass

The actual specific mass of the babassu coconut fibres was determined using the pycnometer method. A certain amount of prepared and dried fibre was weighed. This fibre was placed in the pycnometer and weighed as shown in figure 25. The pycnometer was then filled with water and weighed again. The fibres were immersed for 24 hours. After this period, the same pycnometer was filled only with water and weighed. Then the empty pycnometer was weighed.

The formula for determining the specific mass is as follows:

$$Me = \frac{Mf - Mv}{(Ma - Mv) - (Mfa - \quad)}$$

(Eq. 8)

Where:

$M_e$ = Actual specific mass of the fibres (g/cm$^3$ );

$M_v$ = Mass of the empty pycnometer;

Mf= Mass of pycnometer + fibre;

Mfa= Mass of pycnometer + fibre + water;

$M_a$ = Mass of pycnometer + water.

Figure 25. Mass of pycnometer + fibre.
Source: AUTHOR, 2017.

## 7.2.    2Humidity content

The moisture content was determined when the babassu coconut fibre was exposed to air. To begin the experiment, the fibre was prepared and dried in an oven at a temperature of 60°c until it reached constant mass. The dry mass was then weighed.

Afterwards, the fibre was left exposed for 24 hours, thus obtaining the hygroscopic humidity value.

The moisture content is calculated using the following formula:

$$H = \frac{Msa - Ms}{Ms} \; x \; 100$$

(Eq. 9)

Where:

H= moisture content (%)

$M_{sa}$ = Dry mass exposed to air (g)

$M_s$ = Greenhouse dry mass (g)

### 7.2.3   Water absorption

The following steps were taken to determine water absorption:

The babassu coconut fibre was placed in an oven at 60°C until it reached constant mass, then removed from the oven and immersed in water. For each weighing, the fibre was quickly wrapped in absorbent paper to absorb surface moisture.

Mass determinations were carried out at intervals of 5 min, 30 min, 1 hour, 2 hours and 24 hours until complete saturation was reached.

The following formula was used to calculate water absorption:

$$A = \frac{Mu - Ms}{Ms} \; x \; 100$$

(Eq. 10)

Where:

A= Fibre water absorption (%)

$M_u$ = Wet mass of fibre at a given time (g)

$M_s$ = Greenhouse fibre dry mass (g)

## 7.3   Mould used

The mould was made from 1.5mm thick sheet metal and cut and welded to the dimensions by a locksmith. It has three compartments for making test specimens measuring 19 x 09 x 05 cm per mould, as shown in figure 26.

These dimensions were adopted because they correspond to the most common commercial ecological brick on the market today.

Figure 26. Mould for making bricks. Source: AUTHOR, 2017.

## 7.4    Mix dosing and brick moulding

Soil cement is defined as a hardened product formed by the curing of a compact mixture of soil, cement and water.

For this procedure, the soil complied with the criteria described in standard 8492 (ABNT, 2012), which defined that all material used

should pass 100 per cent through the 4.8mm mesh sieve. After separating the soil, cement and fibre from the babassu coconut epicarp, these elements were weighed out in predefined quantities. After this, the compost was homogenised. For each mix, the dosage began with the mixture of soil and cement. Water was then gradually added. For those mixes in which babassu coconut fibre was incorporated, it was added together with the cement.

The materials were joined together in a large, previously cleaned basin. The entire mixing and moulding process was carried out manually and pressing was done using specific equipment.

It should be noted that during the first stage of dosing (mixing soil with cement), the elements were combined in such a way as to achieve an appearance similar to a damp flour, as shown in figure 27. This process consisted of the reference mix methodology, i.e. without the addition of natural fibre.

48

Figure 27. Mixing soil, cement and water to the consistency of wet farofa.

Source: AUTHOR, 2017.

For the soil-cement specimens with fibres, the predetermined amount of fibres was added to the soil and cement, as shown in figure 28.

Figure 28. Mixture of soil, cement and fibre.

Source: AUTHOR, 2017.

The percentages of fibres that were inserted into the soil-cement mixture, as well as the soil-cement ratios, are shown in Table 6:

Table 6. Number of specimens produced

| Percentage of added fibre | Soil: cement ratio | |
|---|---|---|
| | 05:01 | 10:01 |
| 0% | 10 | 10 |
| 1,50% | 10 | 10 |
| 3% | 10 | 10 |
| Total CP's | 30 | 30 |

Source: Author, 2017.

The mixtures were poured into moulds previously greased with vegetable oil and wrapped with cut low-density polyethylene plastic to facilitate demoulding. The excess mixture on the upper side of the mould was removed and then placed in the manual press for densification in the moulds, applying a constant, evenly distributed load of approximately 1.70 MPa, as shown in figure 29.

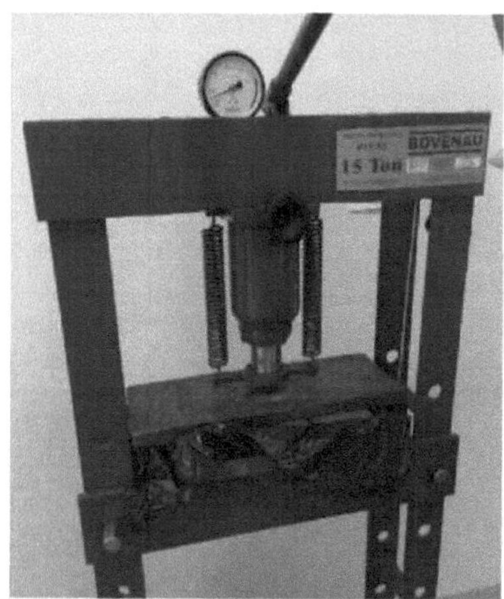

Figure 29. Pressing the ecological brick.
Source: AUTHOR, 2017.

Demoulding was carried out immediately after moulding and the bricks were taken to a room where they were constantly wetted for 7 days, the period needed to cure the cement incorporated into the mixture.

Figure 30. Soil cement bricks 10:01, respectively 0%, 1.5% and 3%.
Source: AUTHOR, 2017.

Figure 31. Soil-cement bricks 05:01, 0%, 1.5% and 3% respectively.

Source: AUTHOR, 2017.

## 7.5 Physical and Mechanical Characterisation of the Composite

The tests for characterising the ecological brick with added babassu coconut fibre will have technical requirements to follow, based on NBR 10834 (ABNT, 2013) as illustrated in table 7:

Table 7. Tests and technical requirements for soil-cement bricks

| Classification | | Observation |
|---|---|---|
| Tolerances Dimensions | Width Height Length | ± 1 mm |
| Compressive strength | Average Values | ≥ 2.0 Mpa |
| | Individual Values | Not less than 1.7 MPa |
| Water absorption | Block | ≥ 20% for average values |
| | | ≥22% for individual values |
| | Brick | ≥ 20% for average values |
| | | ≥ 22% for individual values |

Source: Author, adapted from NBR 10834 (ABNT, 2013)

7.5.1 Water absorption

The water absorption levels of the samples were determined in accordance with NBR 8492 (ABNT, 2012). The level of water absorption could not exceed 20% of the average of its values. The experimental process for obtaining the water absorption content was based on NBR 8492 (ABNT, 2012).

The test was carried out by drying the specimens in an oven at temperatures between 105 °C and 110 °C until they reached the consistency of dough. Afterwards, the brick was allowed to reach room temperature and then weighed in its dry state on a 10kg scale with a sensitivity of 1g. This resulted in the sample mass M1 (g).

The specimens were immersed in a tank for 24 hours. They were then removed from the water tank, where they were superficially wiped dry in order to finally obtain the mass M2 (g) of the saturated brick by weighing it (figure 32).

The values, expressed in percentages, for each sample are expressed using the following expression:

$$A = \left(\frac{M2-M1}{M1}\right) X\ 100$$

(Eq. 11)

In which:

M1 = mass of kiln-dried brick

M2 = mass of saturated brick

A = water absorption, in per cent.

Figure 32. Weighing a brick after immersion in water.
Source: AUTHOR, 2017.

7.5.2 Specific mass

After producing the specimens, each one was weighed and its dimensions measured. In this way, the specific mass of the soil-cement brick can be calculated, knowing that:

$$Me = \frac{Mt}{Vt}$$

(Eq. 9)

Where:

Me= Specific mass;

Mt= Mass of the brick;

Vt= Brick volume.

The values were recorded in a table and then the average was calculated to obtain the specific mass of the ecological brick.


7.5.3 Simple Compression Test


According to NBR 8492 (ABNT, 2012), seven specimens must be prepared. This preparation includes cutting the specimen in half.


perpendicular to the largest dimension and overlapping the faces. In addition, it must be bonded with pre-shrunk cement paste in order to guarantee the time interval for the paste to harden.

The capping should be done with cement paste, which should have a smooth finish and provide parallel and intact tops. The thickness of the paste should vary from 2mm to 3m and the entire procedure should be in accordance with NBR 8492 (ABNT, 2012).After the paste had hardened, the specimens were immersed in water for 6 hours. They were then removed from the water, wiped dry and then placed centrally on the bottom plate of the compression test machine. The simple compression test was then carried out by applying a uniform load, gradually increased at a rate of 50 Kg/s, until a visible crack appeared in the specimen, as shown in figure 33.

Figure 33. Specimen after the simple compression test.
Source: AUTHOR, 2017.

The values for each sample, expressed in MPa, were obtained by dividing the maximum load applied during the test by the area of the sample.

cross-section of the brick. As shown by Equation 12 and in accordance with NBR 8492 (ABNT, 2012).

$$\sigma = \frac{F}{S}$$
(Eq.

Where:

$\sigma$= Simple compressive strength (Mpa);

F = Breaking load (N);

S = Load application area (mm )$^2$

# CHAPTER 8

## RESULTS AND DISCUSSION

### 8.1    Soil characterisation tests

The sedimentation and granulometry test is important to characterise the material so that the cement reacts in the best way and at the lowest cost. For this purpose, according to NBR 8492 (ABNT, 2012), soil that passes 100% of the 4.8mm sieve and contains 10-50% of all the soil passing through the 0.075mm sieve is recommended. The larger graduations serve as filler, while the smaller ones react chemically between the soil and the cement. Figure 34 shows the characteristics of the soil used.

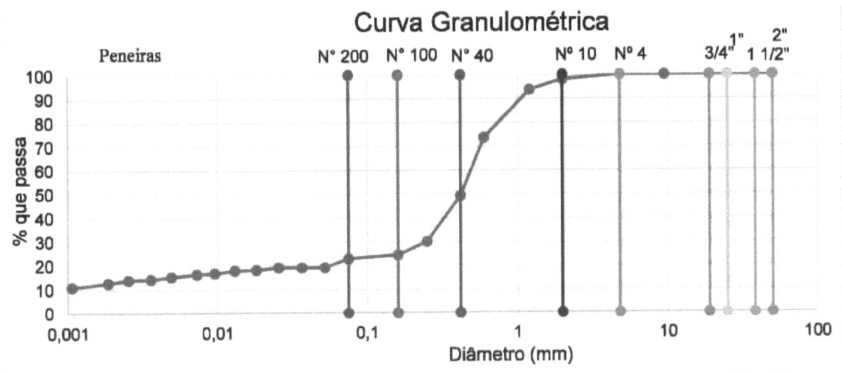


Figure 34. Particle size curve of the analysed soil.

Source: AUTHOR, 2017.

The grain size curve characterises the soil used. Since it is identified as clayey sand, it is ideal for use in soil cement bricks.

### 8.2L  and LP test

The liquidity limit (LL) of the soil was 20.80 per cent, figure 35. The plasticity limit (LP) was 12.27%.

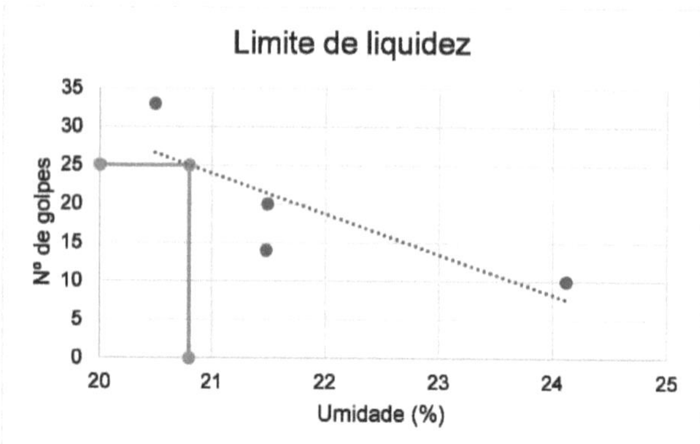


Figure 35. Graph for obtaining the Liquidity Limit of the soil.
Source: Author 2017.

The plasticity index (PI) of the sample was 8.52 per cent, obtained from the numerical difference between the liquidity and plasticity limits. It is therefore identified as a moderately plastic soil.

The value of the soil's natural moisture content, 12.90%, was obtained from NBR 6457 (ABNT, 2016). Thus, according to Equation 7, the Consistency index of the soil is 0.93. Because of this, it is considered a stiff clay.

## 8.    3Specific mass

Table 8 shows the individual results for the specific mass of the soil, obtained from two samples. The average of the values for the two samples was 2.66 g/cm³ .

Table 8. Soil specific mass.

| Sample 40 | | | Sample 55 | | |
|---|---|---|---|---|---|
| M1 | 63,00 | 9 | M1 | 63,00 | g |
| M2 | 724,92 | g | M2 | 668,71 | g |
| M3 | 687,89 | g | M3 | 632,41 | g |
| Temp. | 22,50 | °C | Temp. | 24,00 | °C |
| dt | 0,9977 | g/cm³ | dt | 0,9973 | g/cm³ |
| ds | 2,70 | g/cm³ | ds | 2,61 | g/cm³ |

Source: Author, 2017.

## 8.    4Babaçu coconut fibre characterisation tests

The results of the characterisation of the babassu coconut epicarp used in the ecological brick are shown in Table 9. One of the results corresponds to the expected low density, as the material occupies a large volume without having much weight, remembering that this value corresponds to the volume without the voids.

Table 9. Parameters of babassu coconut fibre

| Specific mass (g/cm )$^3$ | 1,038 |
| --- | --- |
| Moisture content (%) | 7,335 |

Source: AUTHOR, 2017.

As this is a fibre of natural origin, the high rate of water absorption is a common factor analysed in studies. However, it was surprising that this fibre reached values close to 100%, meaning that its mass almost doubled in numerical values due to water, as shown in Figure 36.

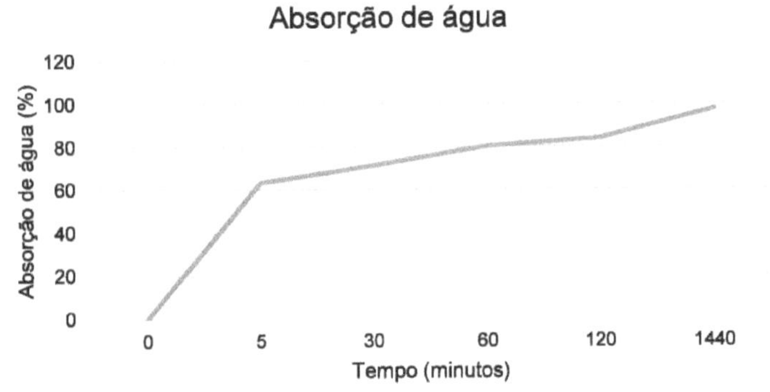


Figure 36. Water absorption of babassu coconut epicarp fibre. Source: AUTHOR, 2017.

This result in water absorption can be detrimental to the brick's strength, as it reduces the water available for cement hydration. In addition, it can influence the decrease in volume, i.e. induce shrinkage of the material.

## 8.    5Physico-mechanical characterisation tests    of the composite

### 8.5.    1Water absorption

Figure 37 and Figure 38 show the water absorption values of the bricks with 0%, 1.5% and 3% fibre additions for the 05:01 and 10:01 mixtures, all according to the curing time and their respective fibre content. The bricks made did not meet the minimum water absorption limit, which must be greater than 20% on average, and the individual value was greater than 22%.

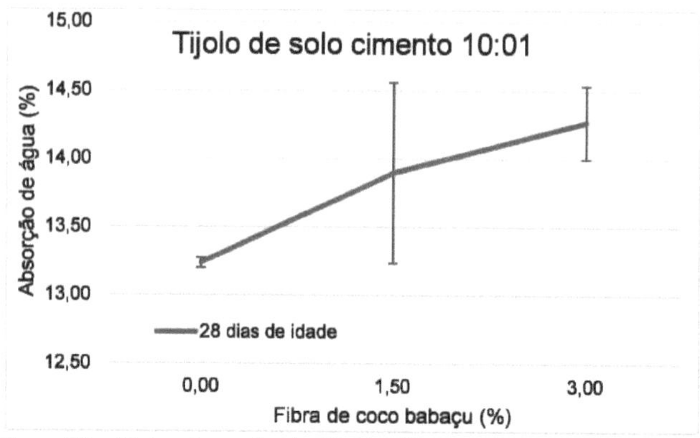


Figura 37.  Water absorption in 10:01 soil-cement brick.

Source: AUTHOR, 2017.

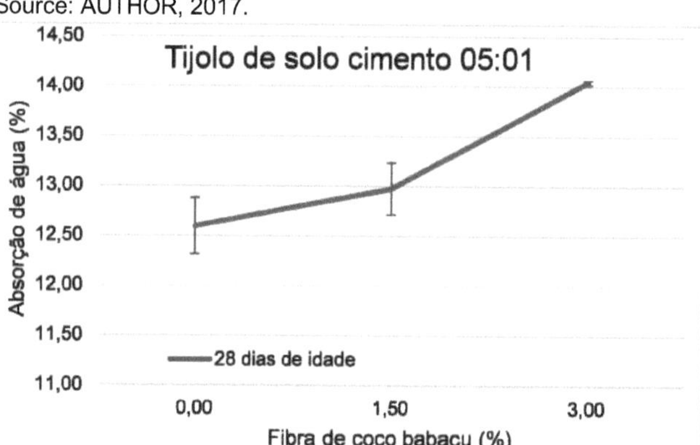


Figure 38. Water absorption in soil-cement brick 05:01.

Source: AUTHOR, 2017.

8.5.    2Specific Mass

By measuring the dimensions of the bricks and their weight, it was possible to know the specific mass for each type of fibre, as shown in figure 39.

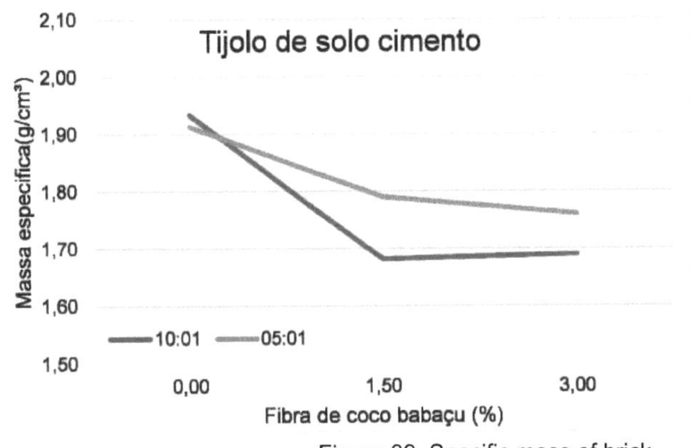


Figure 39. Specific mass of brick.

Source: AUTHOR, 2017.

The values presented show that as the percentage of fibre increases, the specific mass decreases. This is due to the low specific mass of the fibre which, when incorporated, also reduces the mass of the bricks.

## 8.5.  3Simple Compression Test

The simple compression results for the 05:01 and 10:01 traits are shown in Figure 40 and Figure 41, where they are expressed according to the curing time and their respective fibre content.

It can be seen that the bricks with added fibres had higher compressive strength values than conventional soil-cement bricks. At 28 days of age, the 05:01 mix had compressive strength values of 1.80MPa, 2.0MPa and 2.60MPa at 0%, 1.5% and 3% respectively. The 10:1 mix obtained compressive strength values of 0.65MPa, 1.10MPa and 1.50MPa, respectively, at 0%, 1.5% and 3%.

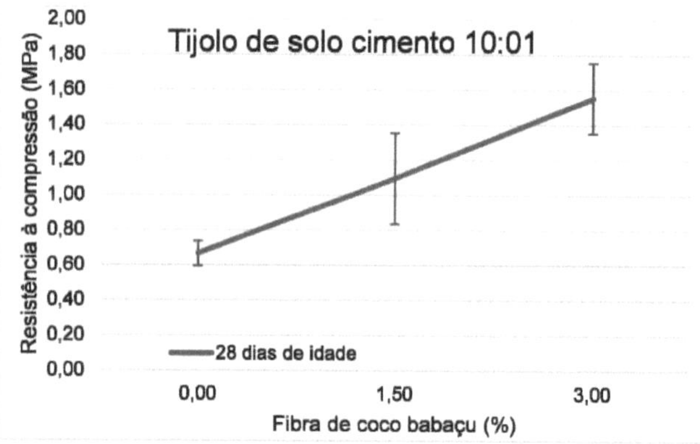



Figure 40. Compression test of 10:01 soil-cement brick.

Source: AUTHOR, 2017.

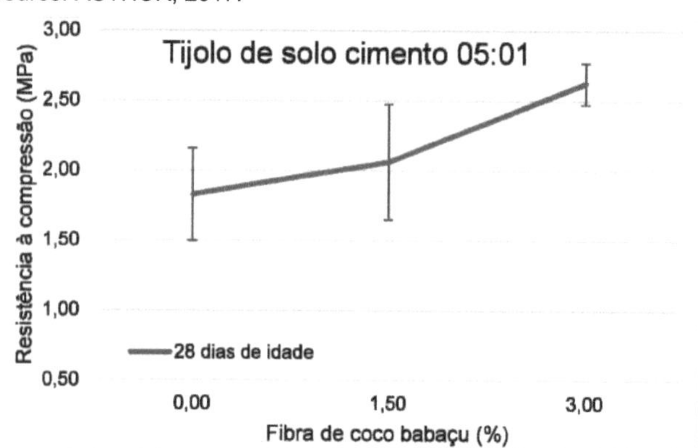



Figure 41. Compression test of soil-cement brick 05:01.

Source: AUTHOR, 2017.

The average of these values contradicts the values required by ABNT standard NBR10834/2013 corresponding to 2.0 MPa. The 05:01 mix with 3% was the one with the best results at 28 days of curing, with 2.60 MPa.

# CHAPTER 9

## CONCLUSION

As presented, the aim of this work is to verify the effects of incorporating different percentages of babassu coconut fibre into ecological soil cement bricks, made from two different designs. Laboratory testing procedures were used to assess the performance of this material. The bricks were analysed for mechanical strength using the compression test, and the specific mass of each brick was also measured. The absorption test was carried out to check the influence of the fibre on the brick.

The addition of babassu coconut fibre to the ecological brick had beneficial results. Within this context, the results were analysed and some considerations were reached.

Incorporating fibre into the brick helped to reduce its specific mass, thus ensuring a lighter material. This factor helps to reduce loads on building structures that use this brick option.

Analysing the mixes used, it was found that the amount of cement is an important factor in determining compressive strength. The mix that showed the best results in the simple compression test was the 05:01 mix; however, it showed lower water absorption and a higher specific mass compared to the 10:01 mix.

However, the addition of babassu coconut fibre to the soil-cement brick also helped to increase the compressive strength. This phenomenon can be understood by the fact that the fibres play a reinforcing role within the composite, as they transfer and redistribute the stresses suffered. The greater the fibre-matrix interaction, the better the result. In both designs, the highest percentage added (3% fibre) represented the best strength obtained.

The addition of fibre to the brick also prevented the sudden rupture and crumbling of the specimens, as it continued to act in the bonding of the elements even after the piece broke.

Another point analysed was the increase in water absorption as fibres were added. This can be explained by the fact that pieces of fibre were exposed in the brick, making it easier for water to enter. The size of the fibre added to the brick also did not follow a standard size or direction for moulding, but was dispersed by incorporation into the matrix. This factor may have influenced the production of voids, since the size of the fibre considerably alters its shape, providing greater curvature. However, the increase in water absorption may have helped the internal curing of the bricks, especially the cement used to make them.

Finally, it should be emphasised that although it produced many satisfactory results, most of them did not meet the requirements set out in standard NBR 10834 (ABNT, 2013). However, this does not detract from classifying babassu coconut fibre incorporated into ecological soil-cement bricks as a qualifier for use in civil construction.

## REFERENCES

ABIKO, Alexandre Kenya. Soil cement: bricks, blocks and monolithic walls. **Construção magazine,** São Paulo, 1983.

ALEXANDRIA, Sandra Selma S. de; LOPES, Wilza Gomes R. In: ENTAC, XI Encontro Nacional de Tecnologia no Ambiente Construído, 2006, Florianópolis. **The use of adobe in the municipality of Uruçuí: A traditional and sustainable construction technique.** Available at: <http://www.infohab.org.br/ entac2014/2006/artigos/ ENTAC2006_ 39353944.pdf>. Accessed on: 14 Apr 2017.

ALVARENGA, Maria Auxiliadora Alfonso. **Design parameters for bioclimatic buildings using adobes in the Araxá region. Evaluation of thermal inertia in adobe and solid brick walls** (Master's dissertation in Architecture). Rio de Janeiro, UFRJ, 1990. Available at: <http://www.proarq. fau.ufrj.br/no vo/trabalhos-de-conclusao/dissertacoes/108>. Accessed on: 14 Apr 2017.

APICER - Portuguese Ceramic Industry Association. **Brick Masonry Manual.** Portuguese Association of Construction Ceramics Manufacturers, Coimbra, 2000.

ARAÚJO, Herbert Gurgel. **Manualisation of Adobe constructions** (Bachelor's

Degree in Civil Engineering). Fortaleza, UFC, 2009. Available at: <http://www.deecc.ufc.br/Download/Projeto_de_Graduacao/2009/Manual izacao %20de%20Construcoes%20em%20Adobe.pdf >. Accessed on 03 Mar. 2017.

BRAZILIAN PORTLAND CEMENT ASSOCIATION - ABCP. **Structural Masonry.** São Paulo, 2010.

**. Manufacture of soil-cement bricks using manual presses.** São Paulo, 1985.

**. Portal Soluções para cidades.** Available at: < http://solucoesparacidades. com.br/habitacao/1-apoio-a-execucao- habitacao/parede-de-concreto/ >. Accessed on: 15 Apr 2017.

BRAZILIAN ASSOCIATION OF TECHNICAL STANDARDS. **NBR 10834:** Soil-cement block with no structural function - Requirement. Rio de Janeiro, 2013.

**. NBR 15270-1:** Ceramic components. Part 1: Ceramic blocks for sealing masonry - Thermology and requirements. Rio de Janeiro, 2005.

**. NBR 15270-1:** Ceramic components. Part 2: Ceramic blocks for sealing masonry - Thermology and requirements. Rio de Janeiro, 2005.

**. NBR 6118:** Design of concrete structures - Procedure. Rio de Janeiro, 2014.

**. NBR 6136:** Simple hollow concrete block for structural masonry. Rio de Janeiro, 2016.

**. NBR 6459:** Soil - Determination of Liquidity Limit - Test Method. Rio de Janeiro, 2016.

**. NBR 6508:** Soil - Specific mass of solids. Rio de Janeiro, 1964.

**. NBR 7180:** Soil - Determination of Plasticity Limit - Test Method. Rio de Janeiro, 2016.

**. NBR 7181:** Soils - Particle size analysis - Test method. Rio de Janeiro, 2016.

. **NBR 8492:** Solid soil-cement brick - Determination of compressive strength and water absorption - Test method. Rio de Janeiro, 2012.

BARBOSA, Elcivone Maria de Lima. **Comparative analysis between ceramic block masonry and drywall.** Revista Especialize, Uberlândia, 2015. Available at:<https://www.google.com.br/url?sa=t&rct=j&q=&esrc=s&source= web&cd=1&cad=rja&uact=8&ved=0ahUKEwjoKj77DTAhVGWpAKHaenA 3UQFggnMA A&url=https%3A%2F%2Fwww.ipog.edu.br%2Fdownload- arquivo- site.sp%3F arquivo%3Delcivone-maria-de-lima-barbosa- 918151415.pdf&usg =AFQjCNFT dOMFQ x6qSS5WX917BZQTIGuvTw&sig2=qyGoMrC15LaAY1E8- vF8qg>.
Accessed on: 9 March 2017.

CARRAZA, Luiz Roberto; SILVA, Mariane Lima da; ÁVILA, João Carlos Cruz. **Technological Manual for the Integral Utilisation of Babassu Fruit.** Institute for Society, Population and Nature, Brasília, 2012.
Available at: < http://www.ispn.org.br/arquivos/Mont_babacu006.pdf>. Accessed on: 06 Mar. 2017

CERRATINGA. **Babassu (Attalea ssp.).** Available at: http://www.cerratinga.org.br /babaçu/. Accessed on: 15 Apr 2017.

BRAZILIAN AGRICULTURAL RESEARCH COMPANY. **Babassu: National Research Programme.** Department of Guidance and Support for Research Programming, Brasilia, 1984. Available at: <https://www. embrapa.br/busca-de- publicacoes/- /publicacao/51263/babacu-programa-nacional-de-pesquisa>. Accessed on: 06 Mar. 2017.

FIGUEIREDO, Davi Messias Corrêa de. **Technical feasibility of the PVC concrete construction system in comparison to the** conventional **masonry system.** (Final Coursework in Civil Engineering). UNIFOR, Formiga, 2015. Available at:<http://bibliotecadigital.uniformg.edu.br: 21015/jspui /bitstream/123456789/294/1/TCC_DaviMessiasCorr%C3%AAadeFigueir edo.pdf>.

Accessed on: 20 March 2017.

FILHO, José Américo Alves Salvador. **Concrete blocks for masonry in industrialised constructions.** 246f. (Doctoral thesis in Structural Engineering). São Carlos, USP, 2007. Available at:< http://www.teses.usp.br theses/available/18/18134/tde-29012009- 104204/en-br.php >. Accessed on: 12 March 2017.

FRANCO, Francisco José Patrício. **Use of Babaçu Coconut Epicarp Fibre in Composites with Epoxy Matrix:** Study of the effect of fibre treatment. (Master's thesis in Materials Engineering). Natal, UFRN, 2010. Available at: <https://repositorio.ufrn.br/jspui/bitstream/123456789/12697/1/ Francisco JPF.pdf>. Accessed on: 06 Mar. 2017.

FURTADO, Fred. **Ecological Brick.** Instituto Ciência Hoje, 2007. Available at: <http://www.cienciahoje.org ,br/revista/materia/id/200/n/tijolo_ecologico>. Accessed on: 14 Apr 2017.

GOLDEN CONCRETE BLOCKS. **Factory of concrete blocks for structural masonry, sealing and closing.** Available at: < http://www. prefabricados deconcreto.com/p/blocos-de-comcreto- fabricante-bh-mg.html >. Accessed on: 12 April 2017.

GRANDE, Fernando Mazzeo. **Manufacture of modular soil-cement bricks by manual pressing with and without the addition of active silica.** (Master's thesis). USP, 2003. Available at: < http://www.teses.usp.br/teses/disponiveis/18/18141/tde-07072003- 160408/en-br.php >. Accessed on 03 Mar. 2017.

NATIONAL INSTITUTE OF METROLOGY, STANDARDISATION AND INDUSTRIAL QUALITY. Consumer Portal. **Ceramic Block (Brick).** 2011. Available at: <http://www.inmetro.gov.br/legislacao/rtac/pdf/RTAC001665.pdf>. Accessed 11

Mar. 2017.

IZQUIERDO, Indara Soto. **The use of natural sisal fibre in concrete blocks for structural masonry.** 128f. (Master's dissertation in structural engineering). São Carlos, USP, 2011. Available at: <http://www.teses.usp.br/teses /disponíveis/18/18134/tde-05042011 - 164738/en-br.php>. Accessed on: 12 Mar. 2017.

KATO, Ricardo Bentes. **Comparison between the Conventional Building System and the Structural Masonry Building System according to the Lean Construction Theory** (Master's Dissertation in Engineering). Florianópolis, UFSC, 2002. Available at: <https://repositorio.ufsc.br/bitstream/handle/ 123456789/111939/193963.pdf?sequence=1>. Accessed on: 03 Mar. 2017.

LIMA, Fabíolla Xavier Rocha Ferreira; SANCHES, Ronaldo Coissi; MARTINS, Vitoria Carvalho. **Soil-Cement Compacted Earth Blocks with Floor and Wall Mortar Residue: Characterisation for Use in Buildings** (Doctoral thesis in Architecture and Urbanism). Brasília, UnB, 2013. Available at: < http: //repositorio.unb.br/bitstream /10482/15550/1/2013_Fab%C3%ADollaXavierRochaFerreiraLima.pdf >. Accessed on: 03 Mar. 2017.

LIMA, Joyce Cavalcante de; SILVA, Fernando Henrique da; SANCHES, Ronaldo Coissi; MARTINS, Vitoria Carvalho; TONETTO, André Luiz; ALVES, Rafael Geraldo. **The Feasibility of Using Soil-Cement Bricks in Civil Construction.** In: Toledo Scientific Initiation Meeting, 2015, Presidente Prudente. Proceedings, Presidente Prudente: Centro Universitário Antônio Eufrásio de Toledo, 2015. Available at <http://intertemas.toledoprudente.edu. br/revista/index.php/ETIC/article/view/5418/5151>. Accessed on: 03 Mar. 2017.

LOOS, Mareio Rodrigo. **Nanoscience and nanotechnology: thermoset composites reinforced with carbon nanotubes.** Rio de Janeiro: Interciência,

2014.

MEDEIROS, Jonas Silvestre. **Unreinforced structural masonry made of concrete blocks:** production of components and design parameters (Master's thesis in Engineering). São Paulo, USP, 1993. Available at: <http://www.pcc.usp.br/files/text/publications/BT_00098.pdf>. Accessed on: 12 Mar. 2017.

MINKE, Gernot: **Construction Manual for earthquake-resistant housing.** 3 ed. Publisher Nordan Comunidad. University of Kassel, Germany, 2005. Available at: https://pt.slideshare.net/miriammorata/manual-de-construo-de-terra- gernot-minke>. Accessed on: 15 Apr 2017.

MONTORO, Paulo. **How to Build Taipa Walls.** Leaflet produced from the "worshop" given by the architect David Easton and his consultancy. Latin American Institute (ILAM), São Paulo, 1994.

MORAIS, Marcelo Brito de. **Feasibility analysis of the application of ecological bricks in contemporary civil construction.** Pensar Engenharia magazine, v.2, n. 2, 2014. Available at: http://revistapensar.com br/ engeharia /pasta_upload/ artigos/a138.pdf>. Accessed on: 09 Mar. 2017

MOTTA, Jessica Campos Soares Silva; MORAIS, Paola Waleska Pereira; ROCHA, Glayce Nayara; TAVARES, Joicimara da Costa, GONÇALVES, Gabrielle Cristina; CHAGAS, Marcela Aleixo; MAGEST, Jalson Luiz;LUCAS, Taiza de Pinho Barroso. **Soil Cement Brick: Analysis of the physical characteristics and economic viability of sustainable construction techniques.** Revista E-xacta, Belo Horizonte,2014. Available at < http://revistas. unibh.br/index.php/dcet/art icle/view/1038 >. Accessed on: 07 Mar. 2017

NASCIMENTO, Cláudio Mario. **Study of coconut fibre as reinforcement in solocement bricks** (Master's dissertation in Mechanical Engineering). Natal, UFRN, 2011. Available at:

<https://repositorio.ufrn.br/jspui/handle/123456789/15675>. Accessed on: 14 Apr 2017.

NASCIMENTO, Maria Victória Leal de Almeida. **Adobe bricks made in Agreste Pernambucano with the addition of shredded tyre rubber.** (Civil Engineering course work proposal). Caruaru, UFPE, 2013. Available at: <https://www.ufpe.br/eccaa/images/documentos/TCC/2013.1/tcc2_ versaofinal201301 %20- 20maria%20victoria%20leal%20de%20almeida%20 nascimento%202.pdf>. Accessed on: 14 Apr 2017.

OLIVEIRA, Célia Ribeiro. **Evaluation of Soil-Cement reinforced with babassu coconut fibres for the production of Ecological Modular Bricks** (TCC presented to the Faculty of Materials Engineering). Federal University of Pará, Marabá, 2011. Available at: < https://femat.unifesspa.edu.br/images/TCCs/2011/TCC-CLELIA- RIBEIRO-DE - OLIVEIRA-2011.pdf >. Accessed on: 08 Mar. 2017

PANZERA, Túlio Hallak. **Development of a ceramic composite material for application in porous bearings.** PhD thesis in Mechanical Engineering. Belo Horizonte, UFMG, 2007. Available at: <http://www.bibliotecadigital. ufmg.br/dspace/bitstream/handle/1843/SBPSB5N3W/monografia_final.p df?sequence=1>. Accessed on: 12 Apr. 2017.

PASTRO, Rodrigo Zambotto. **Structural Masonry Building System** (Undergraduate Monograph in Civil Engineering). UNIVERSIDADE SÃO FRANCISCO, Itatiba, 2007. Available at:<http://lyceumonline.usf.edu.br/sa lavirtual/documentos/1060.pdf>. Accessed on: 20 Mar. 2017.

PESTA, Eloi; MASCARENHAS, Kindersley; PINHEIRO, Luís; QUEIROZ, Maikon; SOUSA, Wellington. **Structural Masonry and its historical development:**

**materials and structural systems.** São Luís, IFMA, 2014. Available at: <
https://pt.slideshare.net/ felipelimadacosta/a-alvenaria-estrutural-and-its-historical-
development>. Accessed on: 02 Mar. 2017

PINTO, Lucas Mazzoleni. **The study of soil-cement bricks with added
construction waste.** (TCC in Civil Engineering). Santa Maria, UFSM, 2015.
Available at: < http://coral.ufsm.br/engcivil/images/ PDF/1_2015/
TCC_LUCAS%20PINTO.pdf >Accessed on: 03 Mar. 2017.

PIRES, lima Bernadette Aquino. **The use of ecological bricks as a solution for
building low-income housing.** (Monograph in the Civil Engineering course).
Bahia, UNIFACS, 2OO4.Available at: <https://issuu.com/jefersondelima/docs/
58137192-tcc-ijolos- ecologicos>. Accessed on: 02 Mar. 2017.

PORTO, Thiago Bomjardim; FERNANDES, Danielle Stefane Gualberto. **Basic
course in reinforced concrete: according to NBR 6118/2014.** São Paulo: Oficina
de Textos, 2015. 210 p. Available at:
<http://itpacaraguaina.bv3.digitalpages.com.br/
users/pblications/9788579751875/pages/11>. Accessed on: 20 March 2017.

PRESA, Marcelo Bastos. **Compressive strength and water absorption in soil-
cement bricks** (Agronomy Degree Monograph). Brasília, UnB, 2011. Available at:
<http://bd m.unb.br/bitst ream/10483/1798/1/2 011_Marcello BastosPr esa.pdf>.
Accessed on: 02 Mar. 2017.

PROMPT, Cecilia. **Bioconstruction Course** (Secretariat for Extractivism and
Sustainable Rural Development). MMA, Brasília, 2008. Available at:
<http://www.mma.gov.br/estruturas/sedrpro ecotur/_ publicacao /140 _publicacao1
5012009110921 .pdf>. Accessed on: 15 Apr 2017.


REZENDE, Dan; GUILHERME, Marcos; ALMEIDA, Túlio. **Soil-Cement Brick**
(Bachelor's Degree in Civil Engineering). Ipatinga, Pitágoras College of Ipatinga,

2013. Available at: < http://www.academia.edu/6842035/ TCC_ Tijolo_ Solo_Cimento >. Accessed on: 03 Mar. 2017.

SALA, L. G., **Proposal for Sustainable Housing for University Students.** 2006. 86 f. Final Course Work (Undergraduate Degree in Civil Engineering). Universidade Regional do Noroeste do Estado do Rio Grande do Sul, Ijuí, 2006. Available at: < http://www.projetos.unijui.edu.br/petegc/wp-content/uploads/tccs/tcc- titulos/2 006/Proposta_de _Habitacao_Sustenta vel_para_Estudantes_Universitarios.pdf >. Accessed on: 09 Mar. 2017

SANTOS, Ana Paula Silva dos. **Behaviour of Soil-Cement-Fibre Mixtures Under Confined Compression with Lateral Stress Measurement** (Master's thesis in Engineering). Porto Alegre, UFRGS, 2004. Available at: <http://www.lume.ufrgs.br/bitstream/handle/10183/3920/000450704.pdf7sequen ce=1&locale-attribute=en_BR>. Accessed on: 04 Mar. 2017.

SANTOS, Everton de Brito. **Comparative feasibility study between ceramic block masonry and cast-in-place concrete walls.**

(Course Conclusion Paper, Federal Technological University of Paraná). Campo Mourão, 2013. Available at:
<http://repositorio.roca. utfpr.edu. br/jspui/bitstream/1/1869/1/CM_COECI_ 2013_1_04.pdf>. Accessed on: 9 Mar. 2017

BRAZILIAN SUPPORT SERVICE FOR MICRO AND SMALL ENTERPRISES. **Business ideas: ecological brick factory.**
SEBRAE. Available at: <https://www.ara.fioc ruz.br/bitsteram/iict/47 3 6/2/175.pdf>. Accessed on: 08 Mar. 2017

SIEIRA, Ana Cristina Castro F.; SAYÃO, Alberto S. F. J.. Influence of damage on the response of geogrids subjected to scuffing. **Revista Escola de Minas.** Ouro Preto, v. 61, n. 4, Oct/Dec. 2008. Available at: <http://www.scielo.br/scielo. php?script=sci_arttext&pid=S0370- 44672008000400017>. Accessed on: 12 Apr.

2017.

SILVA, Cláudia Gonçalves Thaumaturgo da. **Concepts and Prejudices related to Raw Earth Constructions** (Master's dissertation in Public Health). National School of Public Health, Oswaldo Cruz Foundation, 2000. Available at: < https://www.arca.fiocruz.br/bitstream/icict/4736/2/175. pdf >. Accessed on: 03 Mar. 2017

SILVA, Sandra Regina da. **Soil-Cement Bricks Reinforced with Wood Sawdust** (Master's thesis in Structural Engineering). Belo Horizonte, UFMG, 2005. Available at: <http://www.bibliotecadigital.ufmg.br/dspace/bitstream/ handle/1843/BUDB-8C5PAL/reinforced_soil_bricks..._sandra_regina_ da_silva.pdf?sequence=1>. Accessed on: 04 Mar. 2017.

SOLOCAP. **Casagrande.** Available at: <http://www.solocap.com.br/detalhe.asp ?idcod =CASAGRANDE>. Accessed on: 15 April 2017.

SOUSA, Hipólito de. **Masonry Constructions.** University of Porto, Faculty of Engineering. FEUP, 2003. Available at: <http://paginas.fe.up.pt/~earpe/conteudos/TPPC/Sebenta.pdf>. Accessed on 11 March 2017.

SOUZA, Pedro Henrique França Silva Lopes de. **The use of reinforced concrete in the interaction between architecture and structure: analysing Brazilian cases.** (Monograph in Civil Engineering). Faculty of Technology and Applied Social Sciences, University Centre of Brasília, Brasília, 2016. Available at: <http://hdl.handle.net/235/9456>. Accessed on: 20 March 2017.

SPECHT, Luciano Pivoto. **Behaviour of Soil-Cement-Fibre Mixtures subjected to Static and Dynamic Loading for Paving.** (Master's thesis in Engineering). Porto Alegre, UFRGS, 2000. Available at: <http://www.lume.ufrgs.br/bitstream/handle/10183/118555/000270022.pd f? sequence=1>. Accessed on: 04 Mar. 2017.

VIEIRA, Amanda. **Analysing the production process of ceramic bricks in the state of Ceará - from raw material extraction to manufacture** (Monograph for the Civil Engineering course at the Federal University of Ceará). Fortaleza, 2009. Available at: <http://www.brasenic.com.br/bigot/bigot_2.pdf>. Accessed on: 11 March 2017.

ZOROWICH, Ana Clara. **Taipa de Pilão.** Ecoeficientes - Architecture office specialising in Sustainability. Available at: < http://www.ecoeficientes. com.br/taipa-de-pilao/>. Accessed on: 15 Apr 2017.

# I want morebooks!

Buy your books fast and straightforward online - at one of world's fastest growing online book stores! Environmentally sound due to Print-on-Demand technologies.

Buy your books online at
**www.morebooks.shop**

Kaufen Sie Ihre Bücher schnell und unkompliziert online – auf einer der am schnellsten wachsenden Buchhandelsplattformen weltweit! Dank Print-On-Demand umwelt- und ressourcenschonend produziert.

Bücher schneller online kaufen
**www.morebooks.shop**